Discovering Mathematics with the TI-73:
Activities for Grades 5 and 6

Edited and Revised by Melissa Nast

Important notice regarding book materials

Texas Instruments makes no warranty, either expressed or implied, including but not limited to any implied warranties of merchantability and fitness for a particular purpose, regarding any programs or book materials and makes such materials available solely on an "as-is" basis. In no event shall Texas Instruments be liable to anyone for special, collateral, incidental, or consequential damages in connection with or arising out of the purchase or use of these materials, and the sole and exclusive liability of Texas Instruments, regardless of the form of action, shall not exceed the purchase price of this book. Moreover, Texas Instruments shall not be liable for any claim of any kind whatsoever against the use of these materials by any other party.

Permission is hereby granted to teachers to reprint or photocopy in classroom, workshop, or seminar quantities the pages or sheets in this work that carry a Texas Instruments copyright notice. These pages are designed to be reproduced by teachers for use in their classes, workshops, or seminars, provided each copy made shows the copyright notice. Such copies may not be sold, and further distribution is expressly prohibited. Except as authorized above, prior written permission must be obtained from Texas Instruments Incorporated to reproduce or transmit this work or portions thereof in any other form or by any other electronic or mechanical means, including any information storage or retrieval system, unless expressly permitted by federal copyright law. Send inquiries to this address:

Texas Instruments Incorporated
7800 Banner Drive, M/S 3918
Dallas, TX 75251

Attention: Manager, Business Services

M&M's® is a registered trademark of Mars, Incorporated.

We invite your comments and suggestions about this book. Call us at **1-800-TI-CARES** or send e-mail to **ti-cares@ti.com**. Also, you can call or send e-mail to request information about other current and future publications from Texas Instruments.

Visit the TI World Wide Web home page. The web address is: **http://www.ti.com/calc/**

Contents

Preface

This book was written to accompany the TI-73 calculator for teachers teaching 5th or 6th grade. The activities are ordered from easiest to more difficult. Each activity contains the main concept being taught, skills used in the activity, materials needed, an objective, step-by-step instructions for teaching the activity, a wrap-up, assessment, and an extension. Throughout the activity are detailed keystrokes on how to use the TI-73 for the activity. There is also an appendix with common keystrokes for your convenience.

I would like to thank these teachers who developed six of the original activities:

Stephen Davies
Pamela Patton Giles
Gary Hanson
Pamela Weber Harris
Rita Janes
Ellen Johnston

Jane Martain
Linda K. McNay
Louise Nutzman
Aletha Paskett
Karen Wilcox

I would also like to thank Patricia Short for her extensive comments on the book, Belinda Moss and Vince Delisi for their help in piloting the activities and for their editing comments, and Christine Browning and Ruth Casey for their comments on the final draft. I would also like to thank Jeanie Anirudhan, Nelah McComsey, Karl Peters, and Gay Riley-Pfund of Texas Instruments for their assistance in the production of this book.

I hope you and your students enjoy the activities in this book as you explore the wonderful features of the TI-73.

— *Melissa Nast*

About the Author

MELISSA NAST graduated with honors from Texas A & M with a degree in Elementary Mathematics. She has taught Math and Science to 5th and 6th graders for several years. She is currently an instructor for T[3] (*Teachers Teaching with Technology*), teaching other teachers how to use calculators in their classroom. Mrs. Nast resides in Arlington, Texas with her husband and son.

EXPLORATIONS

Activity 1

What's Your Address?

Some lost 4's don't know where their homes are. In this activity, students will use the number 4 and the order of operations to help the 4's find their ten addresses.

Concept
♦ Number sense

Skill
♦ Order of operations

Materials
♦ Student Worksheets (page 3)
♦ TI-73 calculators
♦ Several large card stock pieces with the number **4** written on them
♦ 10 small houses numbered 1 through 10
♦ Rubber stamp and stamp pad

Activity

Post some large 4's on the board with space between each one. (The number of 4's used does not matter.) Put the small houses in a row below the fours. The houses are numbered 1 to 10 and the number on each house represents its address.

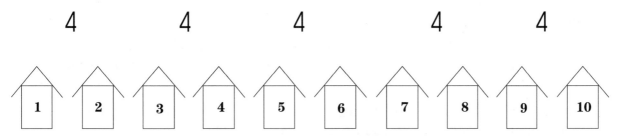

To deliver the 4's to the correct address, instruct the students to use only 4's in equations using any of the operations: addition, subtraction, multiplication, and division. Challenge them to use four 4's if possible. For every address reached using only four 4's, students will earn a stamp. (You could use a rubber stamp.)

Tip: *This activity could be left up year round, with the numbers changing throughout the year.*

Examples:

$4 \div 4 \times 4 \div 4 = 1$

$4 \div 4 + 4 \div 4 = 2$

$4 - 4 \div 4 = 3$

4 [+] 4 [−] 4 = 4

4 [×] 4 [÷] 4 [+] 4 [÷] 4 = 5

4 [÷] 4 [+] 4 [÷] 4 [+] 4 = 6

4 [−] 4 [÷] 4 [+] 4 = 7

4 [+] 4 [−] 4 [+] 4 = 8

4 [÷] 4 [+] 4 [+] 4 = 9

4 [×] 4 [−] 4 [−] 4 [÷] 4 [−] 4 [÷] 4 = 10

There are many different combinations of 4's and operations to make these "addresses." The students could use the calculator to experiment with different ways of finding these answers. They must remember order of operations. (The calculator does have order of operations built into it.)

Wrap-Up

Have students discuss which "addresses" were easier to find. Which were possible using only four 4's? Why is it important to have an order of operations?

Assessment

Repeat the activity but tell the students one of the operation keys is broken. See how this affects the "addresses" that can be found.

Extensions

♦ Students may use the square root, power of 4, factorial, fraction, and percent, along with the operation keys to find all the numbers from 0 to 100.

♦ You could also limit the students to using only four 4's to find all the numbers from 0 to 100. (Some are impossible this way.)

♦ Try using numbers other than 4.

Activity 1

What's Your Address?

Using only fours, find as many expressions as you can that will produce the following answers.

1 =

2 =

3 =

4 =

5 =

6 =

7 =

8 =

9 =

10 =

Activity 2

Factors Galore A:
Common Factors

Students will use the calculator to simplify fractions and investigate common factors and the greatest common factor of the numerator and denominator.

Concept

♦   Number sense/numeration

Skill

♦   Finding common factors, greatest common factor

♦   Simplifying fractions

♦   Prime factorization

♦   Calculator skills: Mode, b/c , SIMP

Materials

♦   Student Worksheets (page 8)

♦   TI-73 calculators

Activity

Before doing this activity, students should have some experience with simplifying fractions, prime numbers, and prime factors.

***Note**: Make sure the calculator is set to **Mansimp** to manually simplify fractions and to **b/c** to input fractions. To do this:*

1.   Press MODE , use the arrows to move down and over to b/c , and press ENTER .

2.   Move to **Mansimp** and press ENTER .

3.   Press 2nd [QUIT] to return to the Home screen.

4.   Have the students enter each fraction from the Student Worksheet into the Home screen of the calculator by typing the numerator, pressing b/c , typing the denominator, and pressing ENTER . The arrow next to the fraction shows that the fraction can be simplified. Discuss with students how they know that the fraction

is unsimplified. For example, in $\frac{18}{24}$, the numerator and denominator are

both even, so we know that 2 is a common factor. Before continuing with the activity, you may want to show examples of other unsimplified fractions on the board and discuss with students how they can be simplified.

5. Press $\boxed{\text{SIMP}}$ $\boxed{\text{ENTER}}$. The calculator displays the new fraction and the lowest prime factor of the numerator and denominator. Record this factor on the recording sheet under **Prime Factors**.

Discuss what the second line on the calculator screen means. (Divide by 2.) On the back of the Student Worksheet, have the students write the arithmetic used to get the simplified fraction. For example, in the window at the right, $\frac{18}{24}$ **Simp Fac=2** would be written as

$\frac{18 \div 2}{24 \div 2} = \frac{9}{12}$. Since we are dividing by 2 (a common factor), $\frac{18}{24} = \frac{9}{12}$.

Continue discussing other lines of simplification.

6. Repeat the $\boxed{\text{SIMP}}$, $\boxed{\text{ENTER}}$, record process until the fraction is completely simplified. (The arrow will disappear.)

Review what a greatest common factor is. (The largest factor that the numerator and denominator have in common.) Guide students to discover that 6 (or 2 x 3) is the greatest common factor of $\frac{18}{24}$. Have them record this under **GCF** on the Student Worksheet. To verify that the original fraction can be simplified using the GCF of 6, enter **18** $\boxed{\text{b/c}}$ **24** $\boxed{\text{SIMP}}$ **6** $\boxed{\text{ENTER}}$.

Wrap-Up

Discuss the process the calculator is using to simplify fractions.

There is a possibility that students will misunderstand the mathematics of division by 1 since the calculator only shows the prime factor and not that the calculator is dividing both numerator and denominator by this prime factor. Students may not understand that they are dividing by 1 in the form of $\frac{GCF}{GCF}$.

Have the students work the problem backwards by taking the simplified fraction and multiplying it by 1 in the form of $\frac{GCF}{GCF}$ to verify that the simplified fraction is equivalent to the original fraction.

Students need to show the mathematics, not just the calculator screen. Encourage them to show that they understand the step-by-step process of how simplification works.

Assessment

Ask students to explain in writing the process that is being used to simplify fractions. Can this be done another way? (Finding the GCF in one step.) Explain.

Extension

◆ Investigate exponents found in the completed table. For example, in the fraction $\frac{128}{640}$, the prime factorization is 2 x 2 x 2 x 2 x 2 x 2 x 2 or 2^7.

Name _____

Date _____

Activity 2

Factors Galore A: Common Factors

Simplify each of the fractions in the table.

FRACTION	PRIME FACTORS	GCF	$F \div \dfrac{GCF}{GCF}$	SIMPLIFIED FRACTION
$\dfrac{18}{24}$	2, 3	6	$\dfrac{18}{24} \div \dfrac{6}{6}$	$\dfrac{3}{4}$
$\dfrac{70}{112}$				
$\dfrac{128}{640}$				
$\dfrac{343}{735}$				
$\dfrac{121}{165}$				
$\dfrac{242}{528}$				
$\dfrac{144}{156}$				
$\dfrac{480}{512}$				
$\dfrac{405}{729}$				
$\dfrac{236}{944}$				
$\dfrac{120}{168}$				
$\dfrac{644}{736}$				
$\dfrac{180}{432}$				

1. Explain what the calculator is doing to simplify these fractions.

    _____

    _____

2. Explain how you found the GCF.

    _____

    _____

Activity 3

Factors Galore B: Get the Low Down

Students practice simplifying fractions in a game format by dividing the numerator and denominator by the greatest common factor.

Concept

- ◆ Number sense
- ◆ Rules of divisibility

Skills

- ◆ Finding common factors
- ◆ Finding greatest common factors
- ◆ Simplifying fractions
- ◆ Mental math
- ◆ Calculator skills: Mode, b/c, SIMP

Materials

- ◆ Student Worksheets (page 13)
- ◆ TI-73 calculators

Activity

Note: *Before beginning this activity, make sure the calculator is set to* **Mansimp**. *To do this:*

1. Press MODE and use the arrows to move down to **b/c** and press ENTER.

2. Press ▼ to move to **Mansimp** and press ENTER.

3. Press 2nd [QUIT] to return to the Home screen.

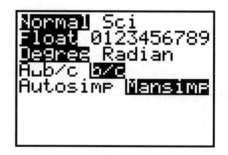

Write an unsimplified fraction (such as $\frac{20}{30}$) on the board, and ask the

students if it is in its simplest form. Guide the students through a discussion of why it is not simplified and how to simplify it. Remind students what a greatest common factor (GCF) is.

Pass out the Student Worksheets. Have students work in pairs. One student predicts the greatest common factor of the numerator and denominator of a fraction and records it in the table. The other student uses the calculator to check this.

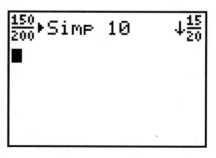

1. Type the numerator, press b/c, type the denominator.

2. Press SIMP, type the guess for the GCF, and press ENTER.

◆ If the arrow disappears, the guess is the greatest common factor and they earn 3 points.

◆ If an arrow does appear, the fraction is not simplified completely, and the student starts over by entering the original fraction and making a new guess. This guess should also be recorded on the Student Worksheet.

◆ Students earn 2 points for getting it right on the second try and 1 point if they get it right on the third try. They should take turns guessing, and after all the fractions are simplified, total their points to determine who has the highest score.

Tip: *If students get the same fraction back when they entered their guess for GCF, they entered a number that was not a common factor. Example:*

$$\frac{150}{200} \blacktriangleright \text{Simp } 7 \qquad \downarrow \frac{150}{200}$$

Wrap-Up

Ask students:

◆ *How did you choose your estimates?*

◆ *What does the calculator do with the guess you entered? (It divides the numerator and denominator by that number.)*

◆ *When did you know your guess was the GCF? (When the arrow disappeared)*

Assessment

Have students answer the questions at the bottom of the Student Worksheet.

Extension

◆ Give the students a new set of numbers and have them try to get the simplified fraction in just one try.

Name _____

Date _____

Activity 3

Factors Galore B: Get the Low Down

Guess the greatest common factor of the numerator and denominator of a fraction and record it in the table. Use the calculator to check your answer. Record your points.

FRACTION	GCF guess #1	GCF guess #2	GCF guess #3	POINTS
$\frac{150}{200}$				
$\frac{48}{16}$				
$\frac{60}{90}$				
$\frac{27}{81}$				
$\frac{16}{48}$				
$\frac{21}{56}$				
$\frac{15}{40}$				
$\frac{22}{273}$				
$\frac{28}{384}$				
			TOTAL POINTS	

1. Why did you choose the numbers you chose?

   _____

2. What does the calculator do with the guess you entered?

   _____

3. When did you know your guess was the GCF?

   _____

4. Write the strategies you used to guess the greatest common factor.

_____

_____

_____

_____

_____

EXPLORATIONS

Activity 4

Factors Galore C: Prime Factorization

Students will use the TI-73 calculator's ability to simplify fractions to find the prime factorization of a number.

Concept

♦ Number sense

Skills

♦ Simplifying fractions

♦ Using prime and composite numbers

♦ Using factors

♦ Prime factorization

♦ Calculator skills: MODE, b/c, SIMP

Materials

♦ Student Worksheets (page 18)

♦ TI-73 calculators

Activity

Note: *Before you begin, make sure the calculator is set to **Mansimp** to manually simplify fractions and to **b/c** to input fractions.*

1. Press MODE, use the arrows to move to **b/c**, and press ENTER.

2. Move to **Mansimp** and press ENTER.

3. Press 2nd [QUIT] to return to the Home screen.

To find the prime factorization of a number, enter the number as the numerator and the denominator of a fraction. (You are entering the number as a fraction to take advantage of the SIMP key on the TI-73. The SIMP key simplifies the fraction one factor at a time.)

For example, to find the prime factorization of 24: Type 24, press b/c, type 24, and press ENTER.

Tip: *Remind students that any number divided by itself is equivalent to 1. You may want to "prove" this to them by dividing 24 by 24.*

The arrow next to the fraction means that the fraction can be simplified. Press [SIMP] [ENTER]. The calculator will display the new fraction simplified by the lowest prime factor. Record this factor on the Student Worksheet. Repeat the [SIMP], [ENTER], record process until the fraction has been simplified completely.

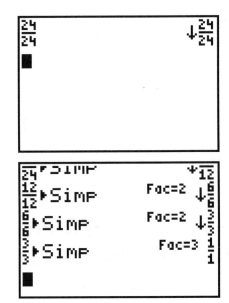

Tip*: For the first two numbers or so, it would be beneficial for the students to do the mathematics on notebook paper to show what the calculator is doing here. 24/24 divided by 2/2 equals 12/12. 12/12 divided by 2/2 equals 6/6 and so on. Discuss with students why the calculator is choosing the factor it is choosing. (It is choosing the lowest common prime factor of numerator and denominator).*

Ask students these questions:

♦ *Will the calculator always display 1 as the last factor? Why or why not?*

♦ *Why don't we record 1 as a prime factor?*

Tip*: The last common factor will be 1. Do not allow students to record this factor because it is not a prime number.*

The product of the common factors is the prime factorization of the number. The factors for 24 will be 2, 2, 2, 3. Students can write this as 2 x 2 x 2 x 3. Ask students if there is a shorter way to write this. (Using powers, 2 x 2 x 2 is 2^3. The prime factorization using powers would be 2^3x 3.)

Wrap-Up

Discuss the process the calculator is using to get the factors. (It is simplifying the fractions by reducing the fraction by the lowest prime factor each time until the fraction is completely simplified.)

Assessment

Explain how you would show that this is the prime factorization of a number. (Make sure students understand the meaning of a prime number and can explain that the product of the prime factors is equal to the number.)

Extension

♦ Show students an alternate way to find the prime factorization of a number by using a factor tree. First, choose any two numbers that you could multiply to get the number. For example, for 24, you could choose 6 and 4. Then do the same for those numbers (the 6 and 4 in this example) until you have all prime numbers.

Example:

```
            24
           /  \
          6    4
         / \  / \
        2  3  2  2
```

The prime factorization is 2 x 3 x 2 x 2.

Prime Factorization Teacher Key

NUMBER	PRIME FACTORS	PRIME FACTORIZATION	PRIME FACTORIZATION (use powers)
24	2,2,2,3	2 x 2 x 2 x 3	$2^3 \times 3^1$
30	2,3,5	2 x 3 x 5	$2^1 \times 3^1 \times 5^1$
36	2,2,3,3	2 x 2 x 3 x 3	$2^2 \times 3^2$
48	2,2,2,2,3	2 x 2 x 2 x 2 x 3	$2^4 \times 3^1$
124	2,2,31	2 x 2 x 31	$2^2 \times 31^1$
128	2,2,2,2,2,2,2	2x2x2x2x2x2x2	2^7
212	2,2,53	2 x 2 x 53	$2^2 \times 53^1$
64	2,2,2,2,2,2	2x2x2x2x2x2	2^6
175	5,5,7	5 x 5 x 7	$5^2 \times 7^1$
420	2,2,3,5,7	2 x 2 x 3 x 5 x 7	$2^2 \times 3^1 \times 5^1 \times 7^1$
844	2,2,211	2 x 2 x 211	$2^2 \times 211^1$
343	7,7,7	7 x 7 x 7	7^3
999	3,3,3,37	3 x 3 x 3 x 37	$3^3 \times 37^1$

Name _____

Date _____

Activity 4

Factors Galore C: Prime Factorization

Use the SIMP *key on your calculator to determine the prime factors for each number. Then complete the table.*

NUMBER	PRIME FACTORS	PRIME FACTORIZATION	PRIME FACTORIZATION (use powers)
24	2,2,2,3	2 x 2 x 2 x 3	$2^3 \times 3^1$
30			
36			
48			
124			
128			
212			
64			
175			
420			
844			
343			
999			

1. What is the calculator doing when you press SIMP ENTER?

   _____

2. Explain how you would prove that this is the prime factorization of a number.

   _____

   _____

   _____

3. Is 2^5 and 5^2 the prime factorization of the same number? Explain why or why not.

_____

_____

_____

_____

Activity 5

Do You See What I See?

Students will discover how pictures formed by graphing ordered pairs can be stretched (and shrunk) by multiplying (and dividing) the coordinates.

Skill

♦   Calculator skills: coordinate graphing, creating lists, WINDOW

Materials

♦   Student Worksheets (page 25)

♦   TI-73 calculators

♦   Transparency of tree graph (page 26)

♦   Copies of tree graph for each student

Activity

Prior to this activity, students should have had some experience with coordinate graphing and understand the difference between x and y axes and how to list an ordered pair.

Show the transparency of the tree to the students. Ask them:

♦   *How can we make the tree taller? (By increasing the y values. For example, multiply each y value by 2.)*

♦   *How can we make the tree shorter? (By decreasing the y values. For example, divide each y value by 2.)*

♦   *What would happen if we doubled the x-coordinates? (The tree would be wider.)*

Give students a copy of the original tree drawing and a Student Worksheet. Have them list the ordered pairs in order, and then have them enter the coordinates in the calculator. This will need to be done in list format. The x-coordinates should be entered in one list and the y-coordinates in a second list. They need to be entered in the order that they would be drawn. (For the calculator to make a connected drawing, students will have to enter ordered pair **A** into the calculator at the beginning and end of the lists.)

To enter items in a list:

1. Clear the Home screen and press LIST.

2. If there are existing lists, press ▶ to move to the first unused list and enter the x-coordinates in order, pressing ENTER after each entry.

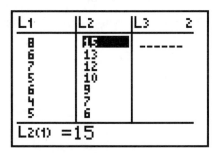

3. Move to the next list and do the same for the *y*-coordinates.

    In the illustration on the previous page, Lists 1 and 2 are used.

To create a plot of the lists:

Tip: *You must have the same number of entries in both* L1 *and* L2 *in order to plot. Otherwise, you will get ERR: DIM MISMATCH when you try to graph.*

1. Before plotting the ordered pairs on the calculator, make sure that only **Plot 1** is on. Press [2nd] [PLOT] and see if Plots 2 and 3 are off. If not, select **4:PlotsOff** and press [ENTER].

2. To graph the tree, press [2nd] [PLOT] and choose **Plot 1** by pressing [ENTER].

3. Select **On** and press [ENTER]. Press [▼] to move to type, use the arrow keys to choose the connected line graph, and press [ENTER]. Move down to **Xlist** and choose the list that your *x*-coordinate data is in by pressing [2nd] [STAT], moving to your list, and pressing [ENTER]. Do the same for **Ylist**. Move down to **Mark**, choose the third mark (the small dot), and press [ENTER].

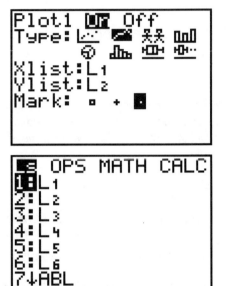

    Tip: *You can use the first mark to make the tree look like it has ornaments on it.*

4. In order to see the entire tree on the calculator, you need to set the viewing window before graphing. Press [WINDOW] and discuss with students what the **Xmin**, **Xmax**, **Ymin**, and **Ymax** values should be.

    Tip: *Do not change the* Δ**X** = *value. The calculator adjusts this value automatically. Leave* **Xscl** *and* **Yscl** *at 1. These determine the distance between tick marks on the graph. To turn off the tick marks, set* **Xscl** *and* **Yscl** *at 0.*

5. Since the highest *x* value is 13 and the highest *y* value is 15, and the coordinates will be doubled, the window settings shown at the right are suggested. We chose large values for **Xmax** and **Ymax** because we want to investigate what happens when we double the values of the coordinates.

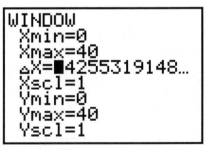

Tip: *If you want everyone's graph to look the same, have all students use the same window values.*

6. Press GRAPH to see the connected
coordinate graph. Use the TRACE key to
explore the graph. Press ▶ repeatedly to
highlight points on the graph in the order
they were entered in the lists. Press ◀ to
move back through the points.

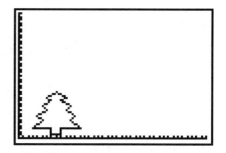

Now the students can begin exploring changing the shape of the tree by
changing the coordinates.

1. To change the coordinates, you need to
get back to the original list that the data
is in. Press LIST and use the arrows to
move to the top of the list containing the
x-coordinates. (Make sure your cursor is
on the list name, not inside the list). If,
for example, you wanted to see what the
tree would look like if the *x*-coordinates
were doubled, press 2nd [STAT], choose **L1**
(the list containing the *x*-coordinates),
multiply by 2, and press ENTER. This will
multiply every *x*-coordinate in **L1** by 2.

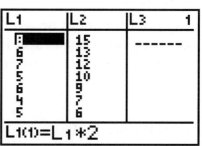

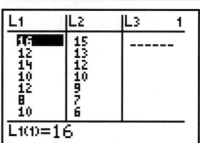

Have students predict and explain what
will happen to the shape.

2. Press GRAPH to see the changes.

Have the students continue to
experiment with changing the shape of
the tree by changing the *y*-coordinates.
Have the students record on the Student
Worksheet how the shape of the tree
changes each time.

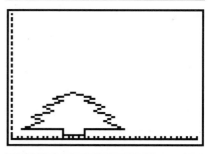

(Press ZOOM **7:ZoomStat** to see the view
shown at the right.)

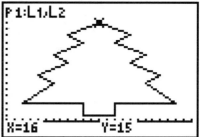

If students want to return to the original
shape, they need to think of how to
"undo" the change they made to the
coordinates. For example, if the values in
L1 have been multiplied by 2, now they
could be divided by 2 to return to the
original values.

Wrap-Up

Ask students these questions:

- *What made the tree tall and narrow? (For example, doubling the x-coordinates but leaving the y-coordinates.)*

- *What made the tree short?*

- *What made the tree wider?*

Assessment

Use another set of ordered pairs to create a picture on the calculator. Have students alter the graph in some way and then give it to a partner to see if the partner can determine how they changed the graph.

Extension

- Have students make a drawing of themselves on graph paper using ordered pairs by measuring their arms, legs, trunk, head, and so forth, and scaling it down to graph paper. Then have them explore by shrinking themselves or making themselves taller using the TI-73.

Name _____

Date _____

Activity 5
Do You See What I See?

List the ordered pairs for the original tree.

A (,)   K (,)
B (,)   L (,)
C (,)   M (,)
D (,)   N (,)
E (,)   O (,)
F (,)   P (,)
G (,)   Q (,)
H (,)   R (,)
I (,)   S (,)
J (,)   A (,)

1. Changing the x- and y-coordinates:

   When I _____ the *x* values, the graph changed by _____.

   When I _____ the *y* values, the graph changed by _____.

2. What made the tree tall and narrow?

   _____

3. What made the tree short?

   _____

4. What made the tree wider?

   _____

Tree Graph

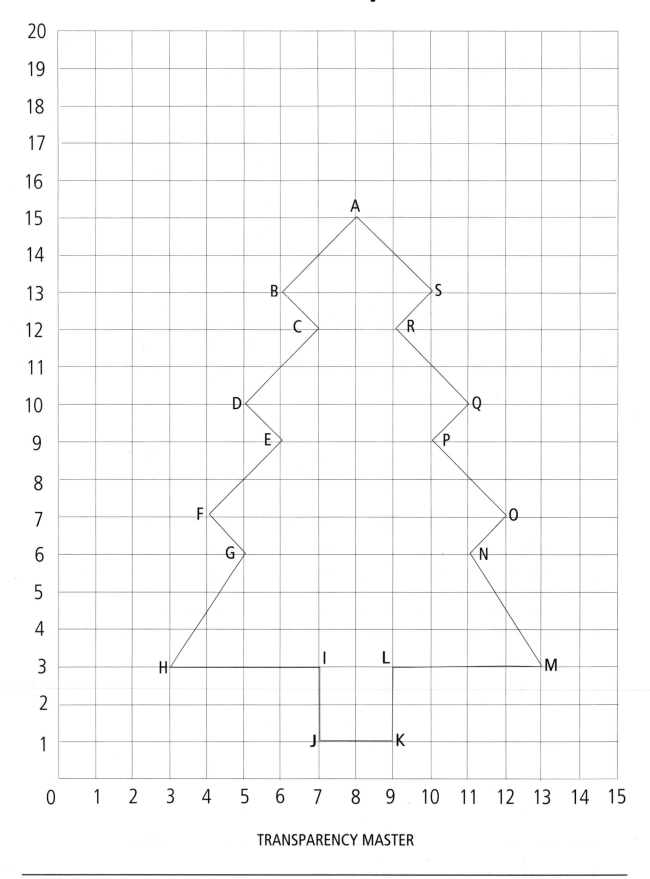

EXPLORATIONS

Activity 6

How Totally Square!

Students will make a variety of sizes of squares using square color tiles. With the squares, the students will discover patterns and be able to make predictions based on the patterns observed.

Activity

Some fifth graders at a local school are planning a banquet for their end-of-the-year party. They are trying to decide how to seat people. Their only option for seating is student desks that measure 1 yard square. They want to arrange the desks in squares to make tables. What are some of their options?

Pass out the color tiles to the students and tell them that they represent the student desks. Have them build square tables with the dimensions of 1*1, 2*2, 3*3, 4*4 until they run out of pattern blocks. Instruct the students to draw the squares on the graph paper (part of the Student Worksheet) as they work. Once they run out of pattern blocks, they can continue drawing the squares on graph paper.

Discuss with the students the patterns they are seeing. How are the total number of blocks related to the side length of each square? Create a table using the information that the students have gathered. (See the table example on the next page.)

Length of side	Total Number of Tiles	
1	1	
2	4	
3	9	
4	16	

Ask the students:

♦ *What do you notice happening to the total number of blocks?*

♦ *Can you see a pattern?*

♦ *What is it? (1*1=1, 2*2=4, and so forth.)*

Guide students to write the pattern using x and y. If the side length is x and the total number of blocks is y, how could we write the pattern using these variables?

$$x * x = y \text{ or } x^2 = y$$

Discuss square numbers with students. What are they? Why do they think they are called "square" numbers? (Point out that all of the shapes built for this activity were squares because the length and width used to build the tables were the same number.)

Show the students how to create a table using the calculator.

1. Press $\boxed{Y=}$.

2. Be sure that the cursor is at the right of the equals sign.

By now your class should have agreed on the formula of $y=x^2$.

3. Press the $\boxed{x}$ key and then the $\boxed{x^2}$ key.

4. Press $\boxed{\text{2nd}}$ [TBLSET] and enter the Table Setup values shown at the right.

5. Press $\boxed{\text{2nd}}$ [TABLE] and an extensive table will display.

Discuss the table with the students.

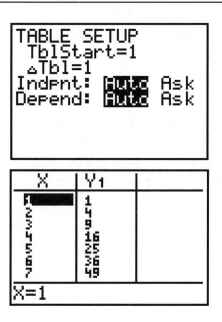

Wrap-Up

Use the table created on the calculator to ask the students these questions:

◆ *How many student desks would be needed to create an 8*8 square table? (64)*

◆ *If there are 121 total student desks, what is the largest size table that can be made? (11x11)*

◆ *Would that be an efficient way to seat every one? (No.)*

◆ *Why or why not? (You would have a lot of "extra" desks in the middle without anyone sitting at them.)*

Assessment

In a math journal, have students write what a square number is and why it is called "square." Have them give at least three examples of square numbers.

Extension

◆ Use the triangular pieces from pattern blocks and investigate triangular numbers and their patterns.

Name _____

Date _____

Activity 6

How Totally Square

Use the color tiles to make squares. Record the length of each side of the square and the total number of tiles in the square in the table below. Look for any patterns that may appear.

Length of Each Side	Number of Tiles

What formula for area did you discover?

EXPLORATIONS

Activity 7

Concept

♦ Measurement

Skills

♦ Finding the perimeter of a square

♦ Patterning

Materials

♦ Student Worksheets (page 34)

♦ TI-73 calculators

♦ Square color tiles

How Totally Square
Part II

*As an extension to Activity 6, **How Totally Square**, students will discover the formula for the perimeter of a square.*

Activity

The fifth grade students planning their end of the year banquet want to know how many students can sit around each square table they create by pushing student desks together.

Pass out the color tiles. Have the students again create squares using the color tiles. (1*1, 2*2, 3*3, 4*4…) Ask them to record on the Student Worksheet how many students can sit around the tables created.

Here is an example of a Student Worksheet.

Length of 1 Side	Number of Students That Can Sit at Table
1	4
2	8
3	12
4	16
5	
6	

Ask the students:

♦ *What do we call the distance around the table or the number of students that can sit around it? (Perimeter)*

- *Do you see a pattern between the number of students that can sit around a table the length of each side of the table? (Yes) What? (The length of a side times 4.)*

- *How would you explain this pattern? What is this pattern? (The formula for the perimeter of a square.)*

- *How many people can sit at a table with a length of 15?*

- *What is the smallest table that can hold 50 people?*

- *Are the values in the **Number of students** column odd or even? Why?*

Help students write the pattern they see as a formula. How can we write our pattern as a formula with variables?

1. We know that the total number of students that can sit at a square table is the length of the side times 4.

2. Let *y* represent the total number of students and *x*, the length of the side of the square. The formula can be written as $y = x * 4$

Have the students create a table on the calculator to investigate this problem further.

1. Press [Y=] and be sure that the cursor is at the right of **Y1**. (Press [CLEAR] if there is already an equation in this spot).

2. Have the students enter the formula they generated (for example, press [x] * 4).

3. Set up the table as shown in the screen at the right.

4. Press [2nd] [TABLE] to see the table format.

Tip: The table will begin with 0. Explain to students that if you have a side that measures 0, then you have 0 square You may want them to start with 1 instead..

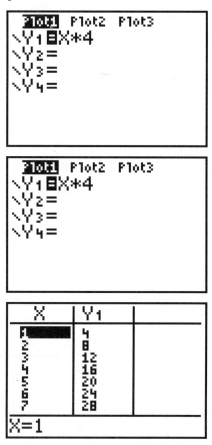

Wrap-Up

Ask the students these questions about the table:

♦ *If there are 110 students in the fifth grade, what are some possible square table arrangements to seat all 110 students?*

Example:

Number of Tables	Dimensions of Table	Number of People per Table	Total Number of People
5	5*5	20	100
1	2*2	8	8
1	1*1	4	4
		TOTAL # OF STUDENTS	112

♦ *How many different arrangements can you find? (Have students record their answers on the Student Worksheet. Explain how close to 110 their answers should be.)*

Assessment

Have students create their own square table arrangements and then have a partner determine the different possible arrangements (similar to the **Wrap-Up** problem). For example, a student may decide to find possible arrangements to seat 60 people. The student would tell his or her partner to find arrangements to seat 60 people and then they would compare answers. They may both have different arrangements, but as long as they are both correct, it is okay.

Describe the table needed to seat your school's basketball team.

Extension

♦ Using your arrangements from the **Wrap-Up**, determine how much money it would cost to cover the tables if material was $2.99 a yard. (Have students calculate the cost to cover only the top of the tables.)

Name _____

Date _____

Activity 7

How Totally Square, Part II

As you use the tiles to create square tables, record your information in this table. Look for any patterns that develop as you work.

Length of Sides	Number of Students Sitting around Table

1. Describe the pattern you see.

   _____

2. Can you describe your pattern as a formula?

   _____

3. What is this a formula for?

   _____

4. If there are 110 students in the fifth grade class, what table arrangements will seat all 110 students? (A few extra chairs are okay.)

Number of Tables	Dimension of Tables	Number of People per Table	Total Number of People
		Total Number of Students	

E X P L O R A T I O N S

Activity 8

A Penny for Your Height

Students will investigate the relationship between inches and centimeters by measuring their height in inches or centimeters and converting the measurement.

Concept

♦ Geometry and measurement

Skills

♦ English to metric conversions

♦ Measuring

♦ Calculator skills: [LIST], mean, [CONVERT]

Materials

♦ Student Worksheets (page 40)

♦ TI-73 calculators

♦ Tape measures (inches and centimeters)

♦ Pennies

Activity

The students from Mount Valley school are participating in a fund raiser for their favorite charity. The students have decided to raise money by donating one penny for each inch of their height. They measured their heights and then challenged their Canadian penpal school to measure their heights and donate money to the same organization. The Canadian penpal students raised a lot more money. Investigate why this happened.

Before beginning this activity, have students look at a ruler and compare a centimeter to an inch. Ask these questions:

♦ *Which is larger, a centimeter or an inch?*

♦ *About how many centimeters fit inside an inch? (About 2.5)*

♦ *Based on this information, who would you expect to raise more money? Why?*

♦ *About how much more money would you expect them to raise? (2 to 3 times more)*

Divide the class into two groups. One group will represent the students from the United States and the others will represent the students from Canada. Try to divide the students evenly by height. (Don't put all the tall students in one group.)

Have the group representing the United States measure their heights in inches. Have the group representing Canada measure their heights in centimeters.

Enter the English measurements in **L1** (list 1) and Metric measurements in **L2** (list 2). Also have students record the data on the Student Worksheet.

For convenience, you may want to clear lists 1 through 4 before you begin the activity.

1. Press [2nd] [STAT].

2. Press [▶] once to select the **OPS** menu.

3. Select **3: ClrList**.

4. To select the lists you want cleared, press [2nd] [STAT] again and select **1: L1**. Repeat this process for **L2, L3,** and **L4** until you have **ClrList L1, L2, L3, L4** on the Home screen.

5. Press [ENTER] to clear the lists.

Now you are ready to enter the data.

1. To begin entering the data in the calculator, press [LIST].

2. Enter the inch heights in **L1**. Press [ENTER] after each list item.

3. Press [▶] and follow the same procedure to enter the centimeter heights in **L2**.

Tip: *You can use any lists you want, but it is beneficial for comparison if the lists are adjacent.*

Ask the students:

♦ *What do you notice about the two lists? (List 2 has larger numbers.) Why?*

Calculate the amount of money each group would collect if they were giving a penny for each unit of measurement in their height. (USA students would give a penny for each inch and Canadian a penny for each centimeter.) Record on the worksheet.

Ask the students:

♦ *Predict which list will have a greater mean. Why?*

Have students calculate the mean of each list and record their answers.

1. Press [2nd] [QUIT] to return to the Home screen.

2. Press [2nd] [STAT]. Press [▶] twice to select the **Math** menu. Select **3: mean(**. This will paste **mean(** on the Home screen.

3. Press [2nd] [STAT], **1:L1** to select list 1.

4. Press [ENTER] to calculate the mean.

5. Repeat the procedure for **L2**, except select **2:L2** to calculate the mean of list 2.

```
Ls OPS MATH CALC
1:min(
2:max(
3:mean(
4:median(
5:mode(
6:stdDev(
```

```
LS OPS MATH CALC
1:L1
2:L2
3:L3
4:L4
5:L5
6:L6
7↓COLOR
```

Compare the means of the two lists.

Ask the students:

♦ *Were your predictions right?*

♦ *Which of the two means is greater? (* **L2** *)*

♦ *Why? (Because the numbers in* **L2** *were larger.)*

♦ *Estimate how many times larger the mean of* **L2** *is than* **L1***? Is that what we should expect? Why? How does this relate to the number of centimeters in an inch?*

♦ *What accounts for the difference in the amount of money each group collected?*

♦ *Can we fairly compare inches with centimeters? How could we change one list so the comparison would be more accurate? (Convert* **L2** *to inches or* **L1** *to centimeters.)*

Before you begin the conversion, be sure that students understand what is happening when they use the conversion option on the TI-73. (When they choose inches to centimeters, the calculator multiplies the values in by **L1** by 2.54.) Select a few students to multiply various heights in inches by 2.54 and share the values they found. Make sure that they say, for example, that 52 inches is about 132 centimeters (132.08 cm.). You may want to have students draw a line segment that is 1 inch in length and measure it with a metric ruler to see that there really are a little more than 2.5 centimeters in 1 inch.

Convert **L1** to centimeters and put the new values in list 3 (**L3**).

1. Press [LIST] and use the arrows to highlight the heading of **L3**.

2. Press [2nd] [STAT] and select **L1**.

3. Press [2nd] [CONVERT] and choose **1:length**. Since we are converting from inches to centimeters, select inches first by pressing **4:inch**.

4. Select **2:cm**. Press [ENTER] and the calculator will convert inches to centimeters.

5. If you would also like to convert **L2** to inches, repeat the process in list 4, or **L4**. Remember to select **2:cm** followed by **4:inch**.

Have students think about the process of conversion. How do you change centimeters to inches? It will be worth the time to ensure that students know and understand what the calculator is doing.

To compare List 2 to List 3, find the mean of List 3. Follow the same procedure as before. First ask the students if they can easily predict which mean will be greater? Why? Then have them compare the actual means. What information does the mean provide about the heights of the students in the two groups?

Wrap-Up

Have each student use his or her own height. Have them calculate the amount of money they would have to give if they were a student from the USA or if they were a Canadian student.

Ask the students:

♦ *If you were in charge of this fund raiser, would you rather have people measure their heights in inches or centimeters? Why?*

♦ *How might you alter the fund raising campaign in the USA (with heights still measured in inches) so that Canadian students would not raise so much more money? (For example, have each person give 2 cents for each inch.)*

Assessment

A student's height is 51 inches. Would this student's height measured in centimeters be a larger number than 51 or smaller? Write a few sentences to explain your answer.

Extension

♦   Investigate other types of measurement such as weight and capacity and compare the difference between metric and standard. Is the ratio always the same as it is between centimeters and inches? Why or why not?

Name _____

Date _____

Activity 8

A Penny for Your Height

Record the heights in inches for the USA students and the heights in centimeters for the Canadian students. Then complete the questions below.

USA students' heights (in inches)			Canadian students' heights (in centimeters)		

Total sum _____inches

Mean _____inches

Amount of money donated to fund raiser (1 penny per inch)

$_____

Total sum _____cm

Mean _____cm

Amount of money donated to fund raiser (1 penny per cm)

$_____

After conversion:
Mean _____inches

1. Why did the Canadian students raise so much more money?

    _____

2. About how many centimeters are in an inch?

    _____

3. How does this relate to the amount of money the Canadian students raised?

    _____

Activity 9

Hey, That's Not Fair! (Or is it?)

Students will use the calculator to simulate dice rolls to play two different games. They will decide if the games give each player an equally likely chance of winning.

Activity

Tip: *You may want to do this activity with real dice first to get students acquainted with probability and dice rolls.*

Ask students:

♦ *What does "fair" mean?*

♦ *Have you ever played a game that you felt wasn't fair?*

♦ *What made it unfair?*

Tell them the definition of fair and unfair games:

♦ When everyone playing a game has an equally likely chance of winning, it is said to be a *fair* game. (For example, if 5 people play, each has a $\frac{1}{5}$ or 20% chance of winning.)

♦ If someone has a clear advantage to win due to the rules of the game, it is an *unfair* game. (For example, if 2 people play, one has a greater than 50% chance of winning.)

Have the students pair up in groups of two. One player will be player A and one will be player B. (They will decide.) They will be playing two different games involving dice rolls on the calculator. They will need one calculator and a Student Worksheet to record the rolls.

*Note: If you have never used the calculators for any random numbers before this activity, seed the random sets in each calculator before you begin by selecting a different number for each calculator and storing it to **rand**. Otherwise, all the calculators will select the same numbers in the same sequence.*

To seed a calculator:

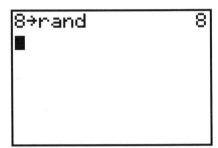

1. Press any number, then press STO▸. (Each calculator should have a different number.)

2. Press MATH, press ▸ to move to **PRB**, select **1:rand**, and press ENTER. (**1:rand** is the default selection.)

Using a ViewScreen unit, show students how to set up the calculator to roll dice.

1. Make sure you are in the Home screen. (If necessary, press 2nd [QUIT] and CLEAR to get a blank screen.)

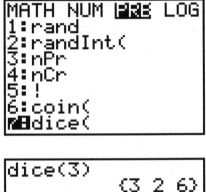

2. Press MATH and press ▸ to move over to **PRB**.

3. Press 7 to choose **dice**. This will paste **dice(** on the Home screen.

4. Since the games require rolling 3 dice, press 3) and ENTER. Three numbers will appear in brackets. These are the three numbers from the first simulated dice roll. Continue pressing ENTER for each roll.

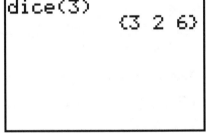

Game 1:

Use the calculator to roll 3 dice at once. If the sum of the three numbers is odd, player **A** gets a point, if the sum is even, player **B** gets a point. Do 20 "tosses" on the calculator. The winner is the player with the most points at the end. Students should record their points on the Student Worksheet.

To verify the validity of games, have students write the 3 numbers and their sum in the appropriate cell. For example, write **3+2+6=11** in the **Player A** column.

Game 2:

Use the calculator to roll all 3 dice at once. Multiply all three numbers. If the answer is odd , player A scores a point and if it is even, player B scores a point. Do 20 "tosses" on the calculator. The winner is the player with the most points in the end. Students should record their points on the Student Worksheet.

Ask the groups:

♦ *Who won game 1, player A or B? (Write the class results on the board in form of a T-chart.)*

♦ *Do you think Game 1 is fair or unfair? Does each player have an equally likely chance of winning?*

♦ *Why or why not?*

♦ *How can we determine if this is a fair game? (By writing out all the possible outcomes or doing a probability tree.)*

♦ *Who won game 2, player A or B? (Write the class results on the board.)*

♦ *Do you think Game 2 is fair or unfair? Why or why not?*

♦ *How can we determine if this is a fair game? (By writing out all the possible outcomes.)*

♦ *How can we list all the possible outcomes? Should we make a chart? Should we use a particular order?*

♦ *What elements should we include in our chart?*

One way to analyze the game to determine fairness is to list all of the possible outcomes. Then you can find the fraction of outcomes that are odd and even. This fraction then is the *probability* of Player A or Player B winning.

You may want to guide the class through the process of writing out all the possible outcomes for Game 1. To list all the possible outcomes, encourage them to use an organized list beginning with a roll of all 1's, and so on. All outcomes need to be considered, for example: 1 1 2, 1 2 1, and 2 1 1 are three different rolls.

Rolls	Sum	Odd/Even
1 1 1	3	Odd
1 1 2	4	Even
1 1 3	5	Odd
1 1 4	6	Even
1 1 5	7	Odd
1 2 2	5	Odd
....		

Students will soon see that there are a lot of outcomes (216 to be exact) and listing them all would be difficult. (If you would like to have all outcomes, divide the task among several groups of students.)

Another way to analyze the game is to think of the individual rolls of the die and whether they are odd or even. Then, look at the sum or product of 3 rolls. An efficient method for organizing the information is a tree diagram. For example:

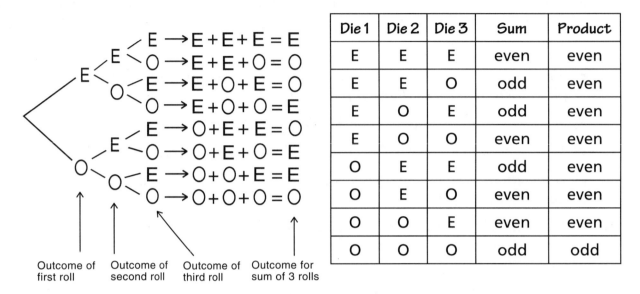

Die 1	Die 2	Die 3	Sum	Product
E	E	E	even	even
E	E	O	odd	even
E	O	E	odd	even
E	O	O	even	even
O	E	E	odd	even
O	E	O	even	even
O	O	E	even	even
O	O	O	odd	odd

Outcome of first roll   Outcome of second roll   Outcome of third roll   Outcome for sum of 3 rolls

The **E**'s represent *even* and the **O**'s represent *odd*. For example, **E + E + E** gives an even sum. So based on the tree diagram (or the accompanying table), $\frac{4}{8}$ of the possible outcomes are even and $\frac{4}{8}$ are odd, making Game 1 a fair game.

A discussion of how students know whether or not the sums are odd or even can be very interesting as students provide their convincing arguments.

Wrap-Up

After students have listed the outcomes for each game, ask:

♦ *How many possible ways are there for you to get an even sum in game 1? (108 if you list all outcomes, 4 if you use a tree diagram.)*

♦ *What about an odd sum? (108 if you list all outcomes, 4 if you use a tree diagram.)*

♦ *Do you think Game 1 is fair? (Yes.)*

♦ *Why or why not? (Same chance of getting an even or odd sum.)*

♦ *How many possible ways are there to get an even answer in game 2? (189 if you list all outcomes, 7 if you use a tree diagram.)*

♦ *Odd answer? (27 if you list all the outcomes, 1 if you use a tree diagram.)*

♦ *Is game 2 fair? (No.)*

♦ *Why or why not? (There is a clear advantage for the player receiving points for even outcomes.)*

Assessment

In a math journal, have students define *equally likely outcomes*. Have them give names of games they have played that were fair and unfair and explain the fairness based on probability.

Extension

♦ Have students create two of their own number games, one fair and one unfair. Encourage them to think of other ways to categorize numbers other than odd or even (for example, divisible by 5, divisible by 10, prime, greater than 6, and so forth). Have them present their games to the class and have classmates play their games to determine the fairness.

Name _____

Date _____

Activity 9

Hey, That's Not Fair

*Use the calculator to roll 3 dice at once. For Game 1, if the sum of the three numbers is odd, player **A** gets a point, if the sum is even, player **B** gets a point. For Game 2, if the product is odd, player **A** gets a point, and if it is even, player **B** gets a point. Do 20 "tosses" on the calculator and record each player's points. The winner is the player with the most points at the end.*

Toss	Game 1		Game 2	
	Player A	Player B	Player A	Player B
1				
2				
3				
4				
5				
6				
7				
8				
9				
10				
11				
12				
13				
14				
15				
16				
17				
18				
19				
20				
Totals				

Tree diagram for Game 1:

Die 1	Die 2	Die 3	Sum

Tree diagram for Game 2:

Die 1	Die 2	Die 3	Sum

Activity 10

Collect All Ten to Win!

Students will simulate a contest that requires the participants to collect 10 different tokens found in the bottom of specially marked cereal boxes. They will determine the average number of cereal boxes a person must buy in order to "collect all 10!"

Concept

♦ Probability

Skills

♦ Collecting and organizing data

♦ Averaging

♦ Calculator skills: MATH, **rand** (generating random numbers)

Materials

♦ Student Worksheets (page 52)

♦ TI-73 calculators

Activity

Ask the students:

♦ *Have you ever tried to collect a set of cards or toys from a fast food restaurant, cereal box, or gum package?*

♦ *How long did it take you to collect all of the set?*

♦ *How much of the product did you have to buy before you collected the entire set?*

Tell them that today they will be simulating a contest from a popular cereal company that requires you to collect 10 pieces in order to win a free box of cereal. Each piece has an equally likely chance of being in the box. Ask the students:

♦ *How many boxes of cereal do you think we will have to buy in order to collect all ten pieces? (Have them write down this prediction on the top of their Student Worksheet.)*

♦ *What is the least number of boxes we will have to buy?*

♦ *How likely is it that we will get all 10 pieces by just buying 10 boxes?*

Use the random integer generator on the calculator to simulate the game pieces in boxes of cereal.

Note: Remember that you must seed the calculators before generating random numbers for the first time. See page 42 for instructions.

```
MATH NUM PRB LOG
1:rand
2:randInt(
3:nPr
4:nCr
5:!
6:coin(
7:dice(
```

```
randInt(1,10)    2
```

1.   Start on the Home screen. Press MATH and ▶ to move to **PRB**.

2.   Select **2:randInt(** and press ENTER to paste to the Home screen.

3.   Enter **1, 10)** since there are 10 game pieces we are trying to collect.

   The calculator will now randomly select numbers between 1 and 10 each time you press ENTER.

Discuss with the students what the random integer generator represents:

♦   Each time they press ENTER, it represents buying a box of cereal.

♦   The number they get when they press ENTER represents the game piece they are getting from that box of cereal.

♦   They will need to continue pressing ENTER and generating random numbers until each number (1 through 10) has appeared at least once.

Students may begin generating the random numbers and recording their data on the Student Worksheet as they go. Here is a sample of a Student Worksheet:

Game Piece	Number of Pieces Acquired
1	II
2	I
3	I
4	II
5	IIIII
6	II
7	I
8	II
9	I
10	I

Total number of boxes the student had to buy: 18 boxes.

Wrap-Up

Once everyone has completed the simulation, compare class data. Ask:

♦ *Did everyone have to purchase the same number of boxes? Why or why not?*

♦ *Who got all ten game pieces by buying just ten boxes?*

♦ *Who thinks they had to buy the greatest number of boxes?*

Find the class average of the number of boxes each person had to "buy." Ask the students:

♦ *What does the class average mean in terms of collecting all ten pieces?*

♦ *How will this affect your shopping in the future?*

Have students complete the remaining questions on the Student Worksheet.

Assessment

Suppose a sticker machine offers six different designs. How many stickers will you have to purchase from the machine in order to collect all six designs?

Extension

♦ Most contests do not give each game piece an equally likely chance. For instance, if the soft drink company Dr. Blue requires contestants to collect game pieces from their soda cans to spell DR BLUE, they may distribute thousands of D's, R's, B's, L's and E's but only a few U's. Students can design a simulation to represent this scenario by using the random integer function again and assigning numbers 1-5 to D, 6-10 to R, 11-15 to B, 16 -20 for L, and 21-25 for E. But only 26 for U. This means that the only time you get the letter U is when 26 comes up on the calculator.

Activity 10

Collect All Ten to Win

Record the results of your calculator simulation of the contest below. Place a tally mark inside the Frequency column each time you get a game piece.

Before you begin, record your prediction:

I predict I will need to buy _____ boxes to collect all ten pieces.

Game Piece	Frequency of getting game piece
1	
2	
3	
4	
5	
6	
7	
8	
9	
10	

1. I had to "buy" _____ boxes to collect all 10 pieces.

2. Class average: _____ game pieces.

3. If our class average represents the average number of boxes a person will need to buy to win, how much money will that person be spending in order to collect all ten pieces to win the coupon? (Each box of cereal costs $3.99.)

    _____

    _____

4. Do you think this contest is worth playing for the coupon? What prize would make it worth playing?

    _____

    _____

Activity 11

EXPLORATIONS

Why Aren't There More Reds in My Bag?

Students will use small bags of M&M's® to make predictions, gather data, and display color distribution results in a circle graph.

Concept

♦ Statistics

Skills

♦ Finding ratios

♦ Creating circle graphs

♦ Sampling data

♦ Understanding sample size

♦ Calculator skills: [TEXT], [LIST], category lists, [STAT]

Materials

♦ Student Worksheets (page 58)

♦ TI-73 calculators

♦ Small bags of Plain M&M's® (47.9g or 21g)

♦ Transparency of circle graph showing actual distribution of colors (page 60)

♦ Manila paper

Activity

Warning: M&M's®, even the plain ones, contain peanut oil. Before doing this activity, check to see if anyone in your class has a peanut allergy. If they do, you may want to use another candy similar to M&M's®.

Distribute the bags of M&M's® to each student with instructions not to open or eat them.

Ask questions about the bags of M&M's®:

♦ *How many M&M's® do you think are in each bag?*

♦ *How many colors do you think are in each bag?*

♦ *How many of each color? Record your predictions on the Student Worksheet.*

♦ *Do you think that each bag has the same colors? Same quantity of M&M's®? Same quantity of each color? Why or why not?*

Have students open the bags and sort the M&M's® by color, again with instructions not to eat.

Tell each student to record the number of each color on the Student Worksheet. Then have them record each color as a ratio, decimal, and percent of the total number of M&M's® in their bag. Discuss the concept of ratio and percent if needed.

Have each student compare the number of M&M's® in his or her bag to those of another student.

Ask students these questions:

♦ *Do you and your partner have the same number of each color of M&Ms®?*

♦ *Does your partner have the same total number of M&M's®?*

♦ *Are all colors represented in both bags?*

Have students arrange the colors of M&M's® in a circle on manila paper, with like colors together and the sides of the M&M's® touching. Trace around the circle. Estimate the center and mark it. Draw a line from the center to the point where the color changes on the circle. Label each section with the color, the ratio to total M&M's®, and percent.

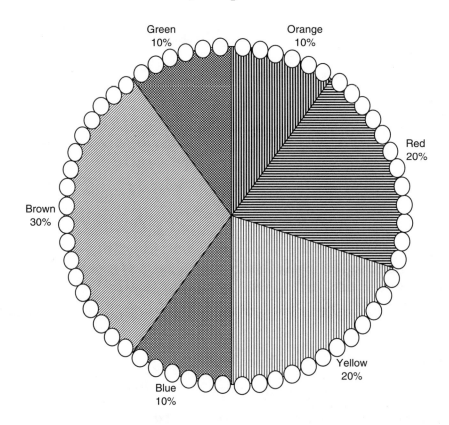

Create a list on the TI-73 called **COLOR**.

1. Press [LIST].

2. Press ▶ to move to the right of **L6** to the first unused list.

3. To name the list **COLOR**, press [2nd] [TEXT] and use the arrow keys to spell **COLOR**, pressing [ENTER] after each letter.

4. When finished, press ▼ to move down to **Done** and press [ENTER].

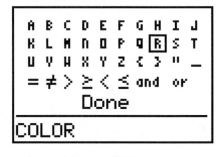

5. Press [ENTER] again and the list will be named.

6. Press [▼] to get **COLOR(1)=**

7. Enter the colors of the M&M's® into the list.

   Be sure to make the list a category list. To do this, enclose the first element in quotation marks (found in the text editor) when you enter it.

8. Press [2nd] [TEXT] and use the arrow keys to spell **BROWN**, pressing [ENTER] after each letter.

9. Press [▼] to move to **Done** and press [ENTER].

10. Press [ENTER] again and **BROWN** will be pasted on the calculator screen under **COLOR**.

11. Continue this process until you have entered all the colors.

Note: You only have to use the quotation marks for the first color entered.

Now you are ready to enter the data.

1. Using the arrow keys, move to the top of list 8 and create a list called **DATA** using the text editor just as before.

2. When finished, press [▼] to get to the first element in the list.

3. Enter the number of each color in this list, pressing [ENTER] after each number.

Tip: This is not a categorical list, so no quotes are needed for the first element.

Create a circle graph from the data.

Note: You do not need to set window values for a circle graph.

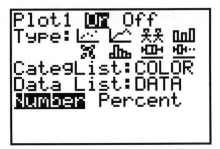

1. On the calculator, press [2nd] [PLOT].
2. Select **1:Plot1** and press [ENTER].
3. Press [ENTER] to highlight **ON**.
4. Press [▼] and [▶] to move to the circle graph and press [ENTER] to select it.
5. For **CategList**, you want the list you named **COLOR**. To get this list, press [2nd] [STAT], use the arrows to move down to **COLOR**, and press [ENTER].
6. Follow the same procedure for the Data List, but choose the list you named **DATA**.

Before graphing, make sure all other plots are turned off.

1. Press [2nd] [PLOT].
2. If a plot other than Plot 1 is turned on, use the arrow keys to highlight that plot and press [ENTER].
3. Use the arrow keys to highlight **Off** and press [ENTER] to turn it off.
4. Press [GRAPH] to see the circle graph.
5. Press [TRACE] and use the arrow keys to explore the circle graph with students.

Have the students sketch the circle graph on the Student Worksheet. Be sure that they label the colors and percents.

Ask the students:

♦ *Compare this graph with the one you created with your M&M's®. What do you notice?*

Combine data with three other classmates by adding the data together and entering it into a new list called **Group**. Create a new circle graph by pressing [2nd] [STAT] and selecting **2:PLOT 2**.

Tip: You cannot display two circle graphs at the same time. You must first turn off Plot 1 and then set this list up as Plot 2.

Sketch and compare this circle graph with the circle graph for your own data. Be sure to label the circle graph.

Repeat the process by combining the whole class data into another list called class. Set this graph up as Plot 3. (Remember to turn off Plots 1 and 2.) Sketch and compare this circle graph with the two previous graphs.

Ask the students:

♦ *How do these three circle graphs compare?*

♦ *How are they different? Why?*

Wrap-Up

Compare the three circle graphs to the circle graph supplied (page 60) showing the actual distribution of colors in a package of plain M&M's®.

♦ *How are the three circle graphs the same?*

♦ *How are they different?*

♦ *Which of your charts is closest to the one showing the actual distribution?*

♦ *Why?*

Note: *At this point you should discuss the idea of a small quantity of data (individual) versus a larger quantity of data (class).*

♦ *Why wouldn't all bags of M&M's® be the same?*

♦ *How do you think they chose the colors for M&M's® that they use?*

♦ *How did they choose how many of each color to include?*

Assessment

Have the students write about why the graphs might be different from others.

Collect the sketches of the circle graphs.

Make a circle graph with different data and give it to students with questions to answer.

Extensions

♦ Do the same experiment using a different candy.

♦ Compare the difference between the color distribution in plain M&M's® bags and peanut M&M's® bags. Is there a difference?

Name _____

Date _____

Activity 11

Why Aren't There More Reds?

Record your predictions in the table below, then open the bags and record the actual number of each color. Determine the ratio, decimal amount, and the percentage of the total number in the bag that each color represents.

Color	Prediction	Actual	Ratio to Total	Decimal	Percent
Brown					
Green					
Yellow					
Red					
Orange					
Blue					

Circle Graph of my data

Circle Graph of my group's combined data

Circle Graph of the class data

Actual Distribution of Colors

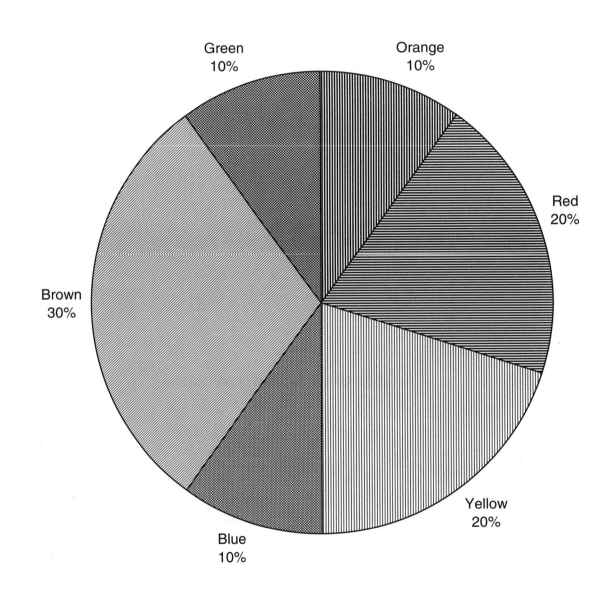

EXPLORATIONS

Activity 12

Gemini Candy

The students will work in cooperative groups to collect data and draw conclusions from random samples simulated by drawing tiles from a bag. The data collected will be represented in the form of a triple bar graph.

Concept

♦ Probability and statistics

Skills

♦ Using fractions

♦ Using decimals

♦ Calculating percent

♦ Collecting and organizing data

♦ Random sampling

♦ Calculator skills: GRAPH , b/c , F↔D , LIST

Materials

♦ Student Worksheets (page 67)

♦ TI-73 calculators

♦ 5 paper bags per group

♦ Color tiles (red, blue and green)

♦ Transparency of **The Problem** (page 68)

Activity

Before using this activity, you should prepare five bags for each group. Label the bags **A**, **B**, **C**, **D,** and **E**, and insert the color tiles into the bags using the table below.

	Red	Blue	Green
Bag A	8	2	0
Bag B	6	3	1
Bag C	4	4	2
Bag D	5	2	3
Bag E	6	3	1

Present students the Problem using the transparency on page 68.

Before placing the students in groups, ask the following guiding questions:

♦ *What two bags do you think belong to the twins?*

♦ *How do you plan to test your guess?*

♦ *How many draws do you need to take from each bag to come to a conclusion?*

♦ *Should every group take the same size sample? Why or why not?*

♦ *What do you need to record and how will you record it?*

Divide the students into cooperative groups of 4 or 5. Pass out the Student Worksheets and guide them as they devise their plan. Then have them do the experiment in their groups, recording the raw data on the Student Worksheet under **Tally**.

After the raw data is collected, instruct student to change it into fractions, decimals and percents for each bag.

Ask the students:

♦ *How this can be done?*

♦ *How can we turn these tally marks into fractions?*
   (Remind students that a probability fraction is $\frac{number\ of\ desired\ outcomes}{total\ number\ of\ outcomes}$ *)*

Example:

Bag A   # of red tiles drawn = 25   # of blue tiles drawn = 5   # of green tiles = 0
         total # of draws   30   total # of draws   30   total # of draws   0

Ask students:

♦ *Can we compare every group's data using only probability fractions? (No, it would be difficult.)*

♦ *Why? (Because not every group did the same number of draws, meaning each group would have a different denominator in their data and it would be difficult to compare data.)*

♦ *How can we solve this problem? (Have each group convert their data to percentages, which can easily be compared.)*

♦ *How can we change these fractions to percents? (Convert to decimals first and then multiply by 100 to make it a percent.)*

Tip: *If you feel your students are not ready for converting fractions to decimals and percentages, the calculator can convert the data into percent for them when they set up the plot to graph. See the note in the Extension for further explanation.*

To convert the fraction to a decimal, use the [F↔D] key on the calculator.

Example: Press 25 [b/c] 30 [F↔D] [ENTER].

To make the data easy to compare with other groups, the students will need to convert the decimals to percents. To convert the decimal to a percent, they will multiply by 100 (or use any other method you have taught them).

A Sample Solution

Bags		Tally Marks	Fraction	Decimal	Percent
A	Red	||||| ||||| ||||| ||||| |||||	$\frac{25}{30}$	.83	83%
	Blue	|||||	$\frac{5}{30}$	.17	17%
	Green		$\frac{0}{30}$	0	0%
B	Red	||||| ||||| |||||	$\frac{15}{30}$	.50	50%
	Blue	||||| ||||	$\frac{9}{30}$	.30	30%
	Green	|||||		$\frac{6}{30}$	.20	20%
C	Red	||||| |||||		$\frac{11}{30}$	.37	37%
	Blue	||||| ||||| |||	$\frac{13}{30}$	.43	43%
	Green	|||||		$\frac{6}{30}$	.20	20%
D	Red	||||| ||||| |||||		$\frac{16}{30}$	.53	53%
	Blue	||||	$\frac{4}{30}$	.13	13%
	Green	||||| |||||	$\frac{10}{30}$	.33	33%
E	Red	||||| ||||| ||||| ||	$\frac{17}{30}$	.57	57%
	Blue	||||| |||	$\frac{8}{30}$	.27	27%
	Green	|||||	$\frac{5}{30}$	.17	17%

Questions to ask the students once they have completed their tables:

♦ *How many samples did your group decide to take? Why?*

♦ *What would have happened if you had taken fewer samples?*

♦ *What does your group's data show?*

♦ *How is your data like or different from other groups?*

♦ *Why did groups get different data if the bag contents were the same?*

♦ *For each bag, what should the total percentage be if you added all the colors? (100 %)*

♦ *Why are some bags 99% or 101%? (Because we rounded off to the nearest hundredths place.)*

Have the students graph their data in triple bar graphs for each bag to determine which bags belong to the twins. They should enter the data in lists.

1. Press [LIST] and press ▶ to move to the right of **L6** to the first unnamed list.

2. To name it **BAGS**, press [2nd] [TEXT] and use the arrow keys to spell **B A G S**, pressing [ENTER] after each letter.

3. When you are finished, highlight **Done** and press [ENTER].

4. Press [ENTER] to paste the name at the top of the list.

Note: To make this a categorical list, the first item entered in the list must be in quotes. For the remaining entries in that list, quotes are not necessary.

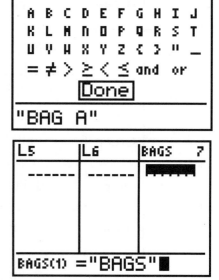

1. Press ▼ once and press [2nd] [TEXT].

2. Use the arrow keys to spell **B A G _ A** (The _ will leave an empty space.)

3. Press ▼ and ▶ to move to **Done** and press [ENTER].

4. Press [ENTER] to paste it on the list.

5. Enter the remaining bags **B** through **E**, using the same process.

6. Press ▶ and ▶ to move to the next list and name it **Red**.

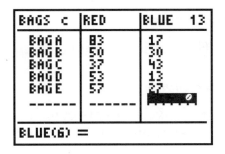

7. Press ▼ and have each group enter their percentage of reds in **Bag A**, **Bag B**, and so forth, pressing [ENTER] after each percentage is entered.

8. Name the next list **Blue** and have each group enter their percentages for blue.

9. Name the last list **Green** and enter the percentages for green.

10. Now make a triple bar graph by pressing [2nd] [PLOT].

11. Make sure all plots are turned off.

12. Choose any plot. Using the arrow keys, set up the plot to look like the screen at the right.

    *Tip: To make the CategList **BAGS**, press [2nd] [STAT] and press the arrows to move down until **BAGS** is highlighted. Press [ENTER]. Do the same to find the appropriate list for DataLists 1, 2, and 3.*

    Note: You do not need to set window values for a bar graph.

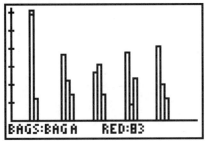

13. Press [GRAPH] to graph the data.

14. Press the [TRACE] key and use the arrows to explore the bar graphs.

Ask the students:

♦ *Which two graphs look the most similar?*

♦ *Based on this, which two bags do you think belong to the twins?*

Tip: If it is difficult for students to choose the graphs that are the most similar, have them look at their data instead.

Wrap-Up

Have students compare their results with the rest of the class. Ask the students:

♦ *Did everyone agree on the two bags that belong to the twins? Why or why not?*

♦ *Did groups that did more draws from the bag have more accurate data?*

♦ *What does the combined class data show?*

Show the students the "real" contents of the bags and discuss the actual data compared to the experimental data.

Assessment

Have students make *Candy Found* posters advertising their results.

Extensions

♦ Have students come up with real life examples in which you might use a sampling procedure similar to the one used in this activity.

♦ Rather than having students calculate percentages, have the calculator do it by entering the raw data (the number of times color was drawn from bag) and selecting **Percent** rather than **Number** when setting up the plots. When you draw a circle graph using the **Percent** option, the calculator will calculate the percentage for each circle part.

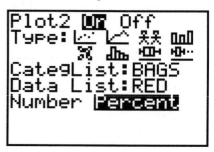

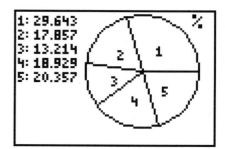

Name _____

Date _____

Activity 12
Gemini Candy

Planning:

1. How will we gather our information?

   _____

   _____

2. How many times should we draw to get a representative sample?

   _____

   _____

3. How will we record our data?

   _____

   _____

Collecting the Data:

Bags		Tally Marks	Fraction	Decimal	Percent
A	Red				
	Blue				
	Green				
B	Red				
	Blue				
	Green				
C	Red				
	Blue				
	Green				
D	Red				
	Blue				
	Green				
E	Red				
	Blue				
	Green				

Gemini Candy - The Problem

Five students went to a candy store on a recent field trip and bought their favorite candy. On the way home, the bags of candy were left on the school bus and the bus driver is trying to determine which bag of candy belongs to which student. The only clue he has is that two of the bags belong to twins and contain the same candy.

Your job is to determine which two bags belong to the twins.

- The candy has been represented by three different colored tiles in five paper bags.

- The bags are labeled A, B, C, D, and E.

- The contents in each bag can only be revealed by pulling out one tile at a time, then replacing the tile in the bag and shaking it before drawing the next tile.

You will design an experiment to determine which two bags have identical contents.

TRANSPARENCY MASTER

Appendix

Commonly-Used Keystrokes

Transparency Masters

This section contains transparency masters for commonly-used keystrokes for the TI-73, including:

♦ Clearing the Home screen

♦ Entering a numerical list

♦ Clearing a list

♦ Clearing all lists

♦ Naming a list

♦ Entering a categorical list

♦ Turning off plots

Clearing the Home screen

- If you have simply been using the number keys, press CLEAR to clear the Home screen.

- If you have been using any of the specialized keys, press 2nd [QUIT] to clear the Home screen.

Entering a numerical list

When creating a list of numerical data, any of the lists can be used.

1. Press LIST to access the lists.

2. Press ▶ and ▲ to move to the list you want to use.

3. Begin entering the data, pressing ENTER after each entry.

```
L1       L2       L3      1
25      ------   ------
63
24

L1(4)=
```

Clearing a list

1. To erase all the data in a list, press ▶ and ▲ to move to the very top of the list.

2. Highlight the name of the list.

3. Press CLEAR ENTER and all the data will be erased.

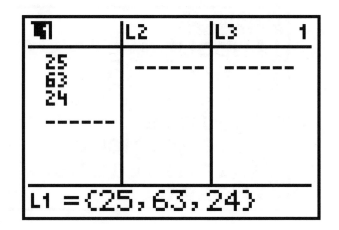

Clearing all lists

To clear all of the lists at one time:

1. Press 2nd [MEM].

2. Choose 6:ClrAllLists.

3. Press ENTER.

Naming a list

If you want a list to have a name other than L1, L2, and so forth:

1. Use the right arrow key to move to the right of Lists 1 through 6 to the first unnamed list.

2. Press ▶ and ▲ to move to the very top of the list.

3. To name it, press [2nd] [TEXT] and use the arrow keys to select the letters for your list name.

4. Press [ENTER] after each entry.

5. Press ▶ and ▲ to move to Done.

6. Press [ENTER] when finished.

7. Press [ENTER] again to paste the list name at the top of the new list.

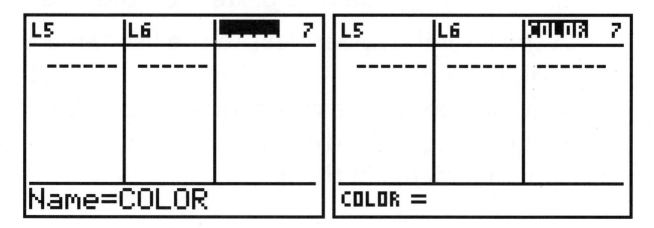

Entering a categorical list

When you want to enter a list of categorical data, such as different colors, use the arrow keys to move past Lists 1 through 6 to the first unnamed list.

1. Once you have named the list, press ▼ once to move down to the first blank element in the list.

2. Press 2nd [TEXT].

3. Use the arrow keys to select ".

4. Press ENTER and continue to use the arrow keys to spell out the first word in the categorical list.

5. Choose the ".

6. Press ▶ and ▲ to move to Done.

7. Press ENTER.

8. Press ENTER again to paste it on the screen.

A small c will appear next to the list name indicating that it is a categorical list. Once the first element in the categorical list is entered, you no longer need to enter quotes for the other elements in the list.

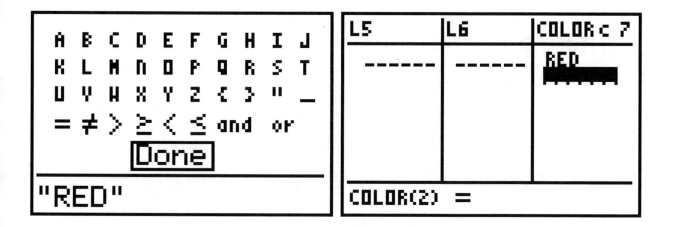

Turning off plots

There are times when you want the calculator to plot only one graph at a time, so it is a good idea to turn off all plots before setting up a new plot to graph.

1. From the Home screen, press [2nd] [PLOT].

2. Choose **4:Plots Off.**

3. Press [ENTER].

4. Press [ENTER] again until it says **Done** on the Home screen.

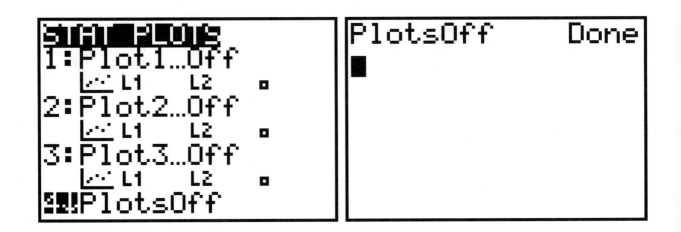

Index

probability, 41, 49, 61, 62

—R—

random integer generator, 50
random numbers
    seeding calculator, 42
random sampling, 61
ratios, 53
rules of divisibility, 11

—S—

sampling data, 53
seeding a calculator, 42
simplifying fractions, 5, 7, 11, 15
square, 31
square numbers, 27, 28
statistics, 41, 53, 61

—T—

table
    creating, 28
TRACE key, 23
tree diagram, 44
tree graph, 26
triple bar graphs, 64
turning off plots, 22, 76 (transparency master)

—W—

window
    setting window values, 22

—Y—

Y= , 28